Practical Guide to FMEA

A Proactive Approach to Failure Analysis

While every precaution has been taken in the preparation of this book, the publisher assumes no responsibility for errors or omissions, or for damages resulting from the use of the information contained herein.

PRACTICAL GUIDE TO FMEA: A PROACTIVE APPROACH TO FAILURE ANALYSIS

First edition. September 27, 2020.

By Mohammed Hamed Ahmed Soliman

Contents

Defining Failure Mode Effect Analysis FMEA and Potential Applications

A n FMEA is a systematic method for identifying and preventing product and process problems before they occur. FMEAs are focused on preventing defects, enhancing safety and increasing customer satisfaction.

FMEAs are conducted in the product design or process development stages, although conducting an FMEA on existing products and processes can also yield substantial benefits.

What is the purpose of a FMEA?
Preventing the process and product problems before they occur is the purpose of Failure Mode Effect Analysis. Used in both the design and manufacturing process, they substantially reduce costs by identifying product and process improvement early in the develop process when changes are relativity easy and inexpensive to make.

FMEA can provide the answer to many problems:
How can we prevent this problem from occurring again in the future?
How can we minimize the risk of this potential failure?
How can we produce an error-free product?

How can we reduce the warranty costs?
How can we improve the safety condition in the workplace?

FMEA as a part of a Comprehensive Quality System
Can FMEA be used alone? While FMEAs can be effectively used alone, a company won't get maximum benefit without systems to support conducting FMEAs.

Two things are necessary needed:

1. A reliable product or process data. Without this data, FMEA becomes a guessing game based on opinions rather than actual facts. Without data the team may focus on the wrong failure modes or missing significant opportunities to improve the failure modes that are the biggest problems.

2. Documentation of procedures. In the absence of documents and procedures, people working in the process could be introducing significant variation in to it by operating it slightly different each time

FMEA is one of the ISO 9001:2000 requirements as you must have a system capable of controlling process that determine the acceptability of your product or services.

Benefits of Failure Modes Effect Analysis "FMEA"
The object of an FMEA is to look for all of the ways a process or product can fail. A product failure occurs when the product does not function as it should or when it malfunctions in some way.

Contribute to improve design for product & process
- Higher reliability.
- Better Quality.
- Increase Safety.

Contribute to cost saving
- Decrease development time & redesign cost.
- Decrease warranty costs.
- Decrease wastes.

Contribute to continuous improvement

FMEA Applies to: System, Process, Design, and Service

FMEA helps manufacturing engineers control the process and eliminate errors during production, thus decreasing warranty costs and wastes.

Service engineers use FMEA to improve the lifecycle of the product and lower its service costs by developing a proper maintenance program.

Potential Applications:
- Equipment components & parts.
- Component proving process.
- Outsourcing/resourcing of product.
- Develop suppliers to achieve quality.
- Major process/ Equipment / Technology Changes.
- Cost Reductions.
- New Product/ Design Analysis.
- Assist in analysis in a flat Pareto chart.

Failure Mode – example to failure modes

A ny event which causes a functional failure. Another definition, ways in which product or process can fail are called failure modes. The FMEA is a way to identify the failures, effects, and risks within a process or product, and then eliminate or reduce them.

Example failure modes:
- Bearing Seized.
- Motor burned out.
- Coupling broken.
- Impeller jammed.

Compressors Failure Modes
Discharge pressure low:
- Air leakage.
- leaking valves.
- Defect gauge.

Engines Failures Mode
Knocking:
 - Pistons hitting the head.
 - Crankshaft plays.
 - Oil pump not function.

Example failure modes, coffee maker

Indeed, even the plain devices have numerous chances for failures. For instance, a trickle espresso producer. A relativity basic family unit machine could have a few things bomb that would deliver the coffeemaker inoperable. Here are a few different ways the espresso make can fizzle:

- The warming component doesn't warm water to adequate temperature to mix espresso.
- The siphon doesn't siphon water into the channel container.
- The espresso producer doesn't turn on consequently by the clock.
- The clock quits working or running excessively quick or excessively moderate.
- There is a short in the electrical rope.
- There is either insufficient or an excessive amount of espresso utilized.

 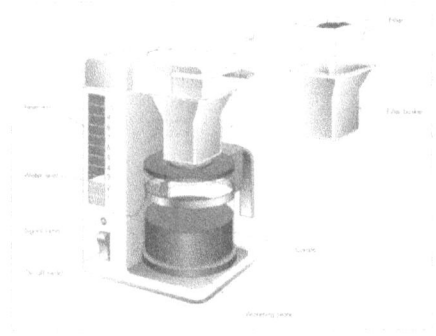

The goal is **100%** Customer Satisfaction

Failures are not limited to problems with the product. Because failures also can occur when the user makes a mistake. Those types of failures should be included in the FMEA. Anything can be done to ensure the product works correctly, regardless of how the user operates it, will move the product closer to 100 percent total customer satisfaction. The use of mistake-proofing techniques, also known by its Japanese term poka-yoke, can be a good tool for preventing failures related to user mistakes.

Failure Effects Description

Local Effect

The failure effect as it applies to the item under analysis.

Ex. Water pump stop.

Next Higher Effect

The failure effect as it applies at the next higher indenture level.

Ex. Water system pressure drop down.

End-Effect

The failure effect at the highest indenture level or total system.

Ex. System stop.

FMEA Team

F MEA is not a job for one individual. The best possible results come when teams are composed of contributors from different engineering perspectives. The team should have between four to six members. Team size is determined by the number of areas affected by the FMEA, for example manufacturing, maintenance, design, engineering, material, technical service, etc. The customer adds another unique perspective and should be considered for team membership. If customers cannot be included, the team should devise ways to generate voice-of-the customer data.

Team Leader:
The team leader is responsible for coordinating the FMEA process as follow:
1. Setting up and facilitate meeting.
2. Ensure that all resources are available.
3. Make sure the team moves toward completing the FMEA process.
4. The team leader role is more like of a facilitator rather than decision maker.
5. Determine the boundaries of freedom.
6. Define the scope of the project.

Steps of FMEA Process

There are twelve steps for a successful FMEA process.

1. Select a high-risk process, then follow these steps.
2. Review the process: this step usually involves a carefully selected team that includes people with various job responsibilities and levels of experiences. The purpose of an FMEA team is to bring a variety of perspectives and experiences to the project.
3. Breakdown the system into components and sub-components.
4. Brainstorm potential failure modes.
5. List potential effects of each failure mode.
6. Assign a severity ranking for each effect.
7. Assign an occurrence ranking for each failure mode.
8. Assign a detection ranking for each failure mode.
9. Calculate the risk priority number (RPN) for each effect.
10. Prioritize the failure modes for action using RPN.
11. Take action to eliminate or reduce the high-risk failure modes.
12. Calculate the resulting RPN as the failure modes are reduced or eliminated.

FMEA Working Sheet

Component/Item Name:

Function

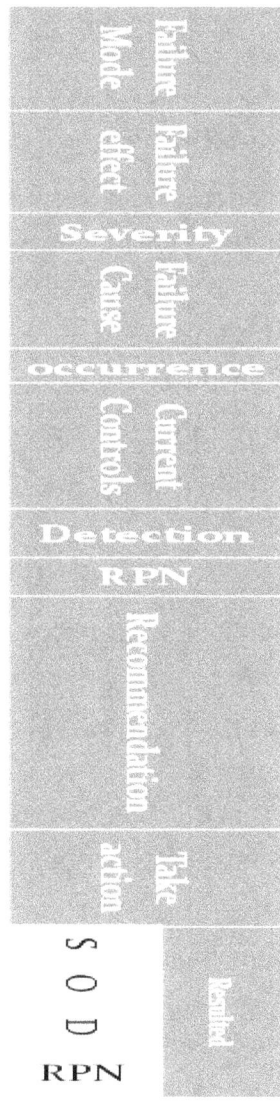

Step.1 Select a critical process/product/equipment by performing a criticality analysis.

The rule of thumb, is safety first! So, if there is a system that the consequence of its failure can have high impact on safety, it should be given high priority attention: eg. Firing system, elevator.

Step.2 Review the Process or Product

If the team is considering a product, they should review the engineering drawing of the product.

If the team considering a process, they should review the operation flowchart.

This is to ensure that everyone has the same understanding about the process or product.

For a product, they should physically see the product and operate it.

For a process, they should physically walk through the process exactly as the process flows.

Step.3 Breakdown the System into Components and Sub-components

If the system is a large system, like a water system that supplies an industrial process, the pump can be a critical component inside the system. A motor pump is a critical subcomponent because its failure can break down the entire process. The motor pump should be broken down into more subcomponents that are likely to fail and will affect the system, such as the motor's bearings and the rotor shaft. The FMEA will be used to prevent the probability of failure for each component or subcomponent.

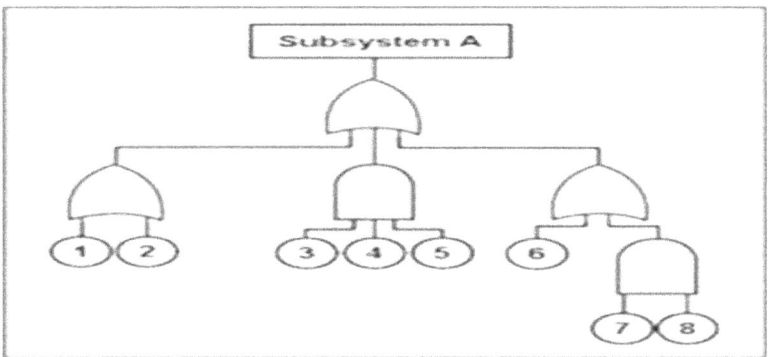

Step.4 Brain Storm Potential Failure Modes

When everybody in the group has a comprehension about the item or the cycle, colleagues should start contemplating the potential failure modes that could influence the process or the item quality.

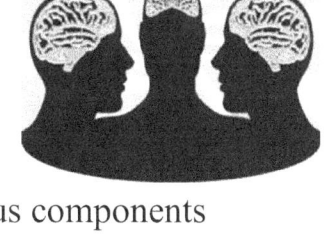

Zeroing in ought to be on the various components (individuals, material, management, strategy… and so on). When the conceptualizing is finished, the thoughts ought to be sorted out by gathering them into like classes. There are numerous approaches to assemble failure modes, they can be gathered by kind of failure (electrical, mechanical, client made). Where on the item or cycle the failure happens.

Main Rules of Brainstorm:

Try not to remark on, judge or study thoughts at the time they are advertised. Empower inventive and odd thoughts. The objective is to wind up with an enormous number of thoughts; and assess thoughts later. Every thought ought to be recorded and numbered precisely as offered, on a flip outline.

Expect to generate at least 50 to 60 concepts in a 30-minute brainstorming session.

Failure Mode & Effect Analysis FMEA

Item	Function	Failure Mode	Failure effect	Severity	Failure Cause	occurrence	Current Controls	Detection	RPN	Recommendation	Take action	Resulted			
												S	O	D	RPN

-How can this sub system fail to perform its function?
-The Way the failure occurred
-What will the operator see?

Step.5 List Potential Effects for Each Failure Mode
For a portion of the failure modes, there might be one impact, while for different modes, there might be a few impacts. This data must be through that it will take care of into the task of the risk ranking for every one of the failures.

Tips:
One failure mode could have several effects. For example, an electrical cutoff in the home could stop the refrigerator and damage food or prevent you from doing work on the computer.

Several failure modes could have one effect. A dead car battery or tire failure has the same effect on your vehicle – it will be difficult to make it to work on time with such a failure early in the morning.

The team must determine the end-effect each failure mode has on the system or the process. This means examining how each failure affects the entire system, the facility or the other connected processes.

Failure Mode & Effect Analysis FMEA

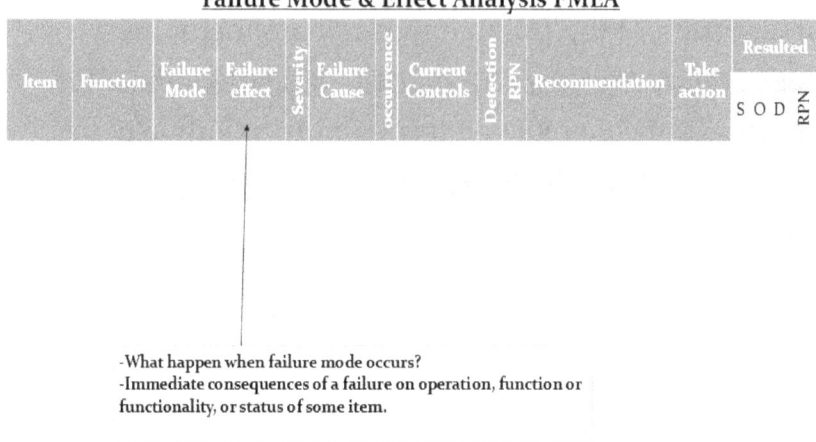

-What happen when failure mode occurs?
-Immediate consequences of a failure on operation, function or functionality, or status of some item.

Steps 6-8 Assign Severity, Occurrence, and Detection Rankings

Each of these three rankings is based on 10-point scale, with 1 being the lowest ranking, and 10 the highest.

Failure Mode & Effect Analysis FMEA

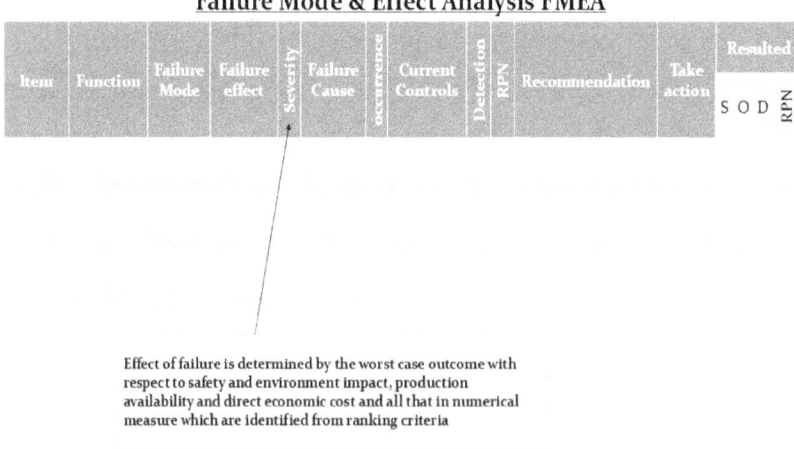

Effect of failure is determined by the worst case outcome with respect to safety and environment impact, production availability and direct economic cost and all that in numerical measure which are identified from ranking criteria

Failure Mode & Effect Analysis FMEA

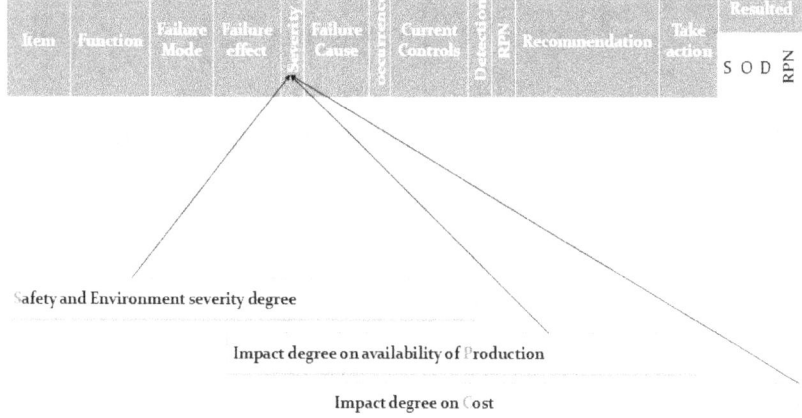

Safety and Environment severity degree

Impact degree on availability of Production

Impact degree on Cost

Severity Ranking Criteria

Description of Failure Effect	Effect	Ranking
No reason to expect failure to have any effect on Safety, Health, Environment or Mission.	None	1
Minor disruption of production. Repair of failure can be accomplished during trouble call.	Very Low	2
Minor disruption of production. Repair of failure may be longer than trouble call but does not delay Mission.	Low	3
Moderate disruption of production. Some portion too of the production process may be delayed.	Low to Moderate	4
Moderate disruption of production. The production process will be delayed.	Moderate	5
Moderate disruption of production. Some portion of production function is lost. Moderate delay in to High restoring function.	Moderate to High	6
High disruption of production. Some portion of production function is lost. Significant delay in restoring function.	High	7
High disruption of production. All of production function is lost. Significant delay in restoring High function.	Very High	8
Potential Safety, Health or Environmental issue. Failure will occur with warning.	Hazard	9
Potential Safety, Health or Environmental issue. Failure will occur without warning.	Hazard	10

Step.7 Assign an Occurrence Ranking for each Failure Mode

The best technique for deciding the occurrence ranking is to utilize real information from the process. This might be as failure history. At the point when real failure information is not accessible, the group must gauge how frequently a failure mode may happen, the group can improve gauge on how likely a failure mode is to happen and at what recurrence by knowing the expected reason for failure. When the potential causes have been distinguished for the entirety of the failure modes, an occurrence ranking can be appointed regardless of whether the failure information are not existed.

Failure Mode & Effect Analysis FMEA

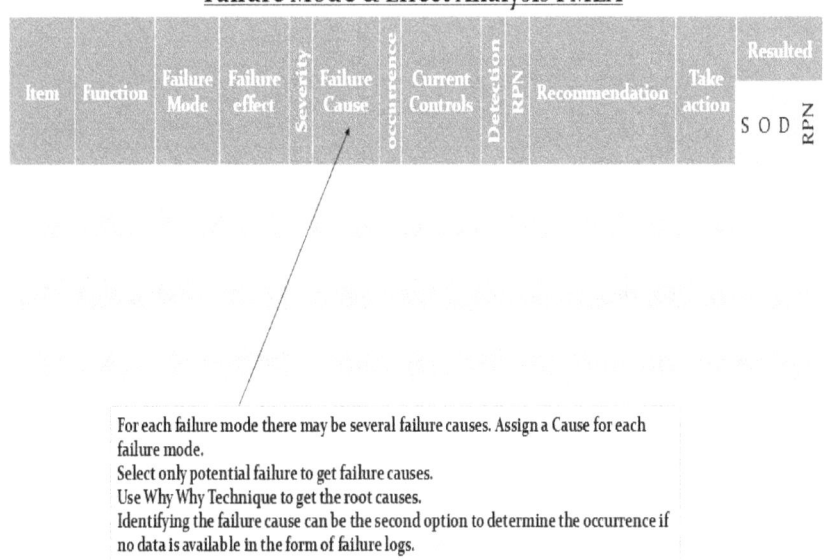

For each failure mode there may be several failure causes. Assign a Cause for each failure mode.
Select only potential failure to get failure causes.
Use Why Why Technique to get the root causes.
Identifying the failure cause can be the second option to determine the occurrence if no data is available in the form of failure logs.

Failure Mode & Effect Analysis FMEA

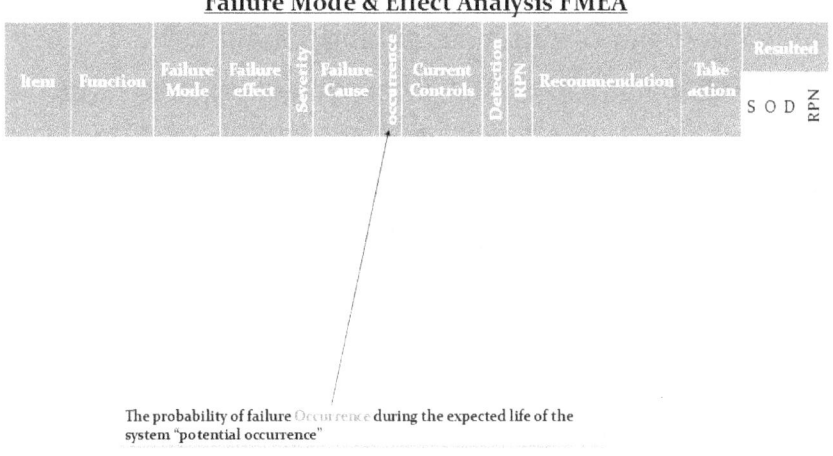

The probability of failure Occurrence during the expected life of the system "potential occurrence"

Occurrence Ranking Criteria

Rank	Freq	Description
1	1/10,000	Remote probability of occurrence; unreasonable to expect failure to occur
2	1/5,000	Low failure rate; similar to past design that has, in the past, had low failure rates for given volume or load
3	1/2,000	Low failure rate; similar to past design that has, in the past, had low failure rates for given volume or load
4	1/1000	Occasional failure rate; similar to past design that has, in the past, had similar failure rates for given volume or load
5	1/500	Moderate failure rate; similar to past design that has, in the past, had moderate failure rates for given volume or load
6	1/200	Moderate failure rate; similar to past design that has, in the past, had moderate failure rates for given volume or load
7	1/100	High failure rate; similar to past design that has, in the past, had high failure rates that have caused problems
8	1/50	High failure rate; similar to past design that has, in the past, had high failure rates that have caused problems
9	1/20	Very High failure rate; almost certain to cause Problems
10	1/10	Very High failure rate; almost certain to cause Problems

Operating hours based on the automotive industry benchmark.
Ranking can be determined based on historical data or similar system benchmarking

Step.8 Assign a Detection Ranking for each Failure Mode and/or Effect

To start with, the current control ought to be recorded for the entirety of the failure modes, or impacts, and afterward the recognition rankings appointed. In the event that one failure mode or impact has a few causes, recognition and occurrence rankings ought to be relegated dependent on these causes. At the point when potential causes are disposed of, the danger of failure is brought down.

In case the application is for an equipment maintenance, current control methods can be the current preventive maintenance program and or the current detection methods (condition monitoring program).

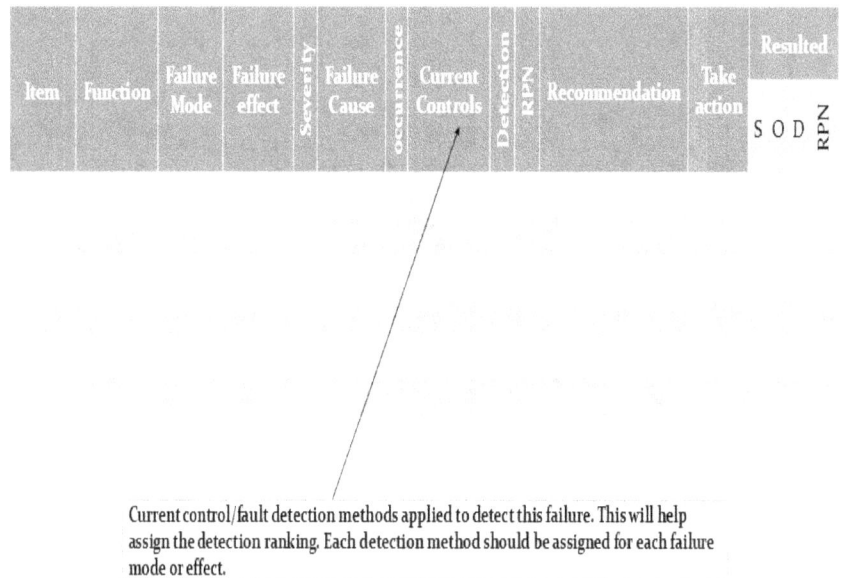

Current control/fault detection methods applied to detect this failure. This will help assign the detection ranking. Each detection method should be assigned for each failure mode or effect.

Failure Mode & Effect Analysis FMEA

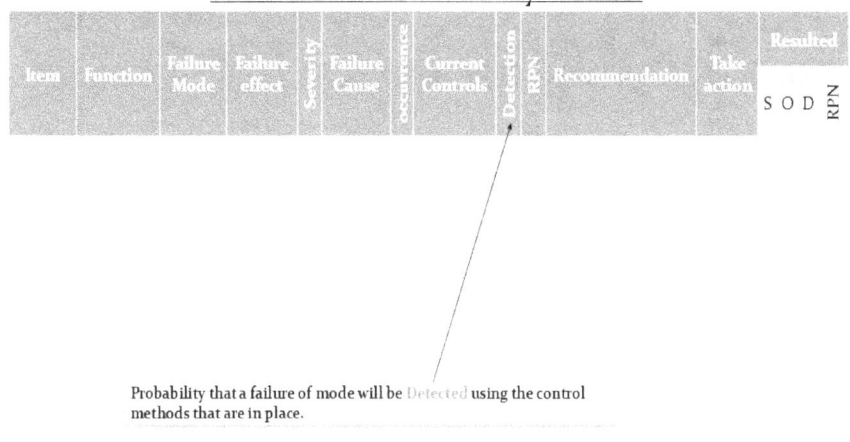

Probability that a failure of mode will be Detected using the control methods that are in place.

Detection Ranking Criteria

Rank	Description
1-2	Very high probability of detection
3-4	High probability of detection
5-7	Moderate probability of detection
8-9	Low probability of detection
10	Very low probability of detection

Step.9 Calculate the Risk Priority Number RPN

Risk Priority number= Severity x Occurrence x Detection.

This number alone is meaningless because each FMEA has a different number of failure modes and effects. However, it can serve as a gauge to compare the revised RPN once the recommended actions have been instituted.

Failure Mode & Effect Analysis FMEA

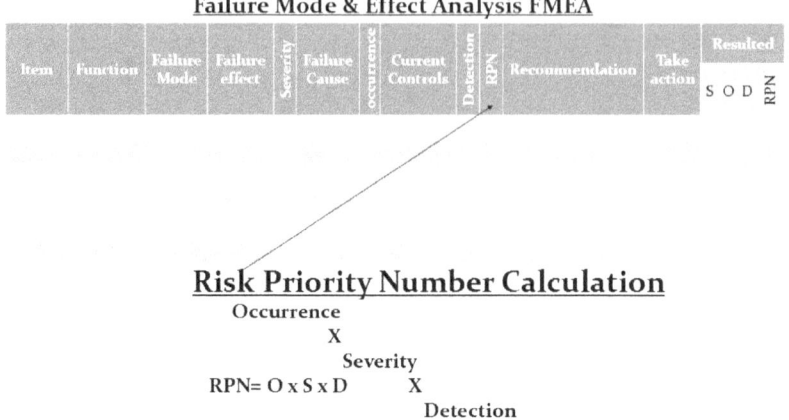

Risk Priority Number Calculation

Occurrence

X

Severity

RPN= O x S x D   X

Detection

What is RPN?

The Risk Priority Number (RPN) methodology is a technique for analyzing the risk associated with potential problems identified during a Failure Mode and Effects Analysis (FMEA).

RPN Calculation Benefits:

- Contribute in Risk Assessment.
- Compare components to determine priority for corrective action. Components with higher RPN are given more attention.
- Assessing the risk priority number.

Each potential failure mode or effect is rated in each of these three factors on a scale ranging from 1 to 10. By multiplying the ranking a risk priority number RPN can be determined for each potential failure mode and effect.

The RPN will range from 1 to 1000 for each failure mode. It is used to rank the need for corrective action. Those failure modes with the highest RPN number should be attended first. Although the special attention should be given when the severity ranking is high from (9 to 10) regardless of the RPN.

Once a corrective action is takes, a new RPN is determined. This new RPN is called the resulting RPN.

Step.10 Prioritize the Failure Modes for Action
Failure modes ought to be organized by positioning them all together, from the most elevated danger need number to the least. Odds are that you will find that the standard 80/20 principle applied with the RPNs. A Pareto chart should be created.

The group should now choose which thing to work for. Typically, it assists with setting a cutoff RPN (cutoff point), where any failure modes with a RPN over that point are taken care of. Those beneath the cutoff are disregarded until further notice.

Tips:

High-risk numbers should be given attention first; then you can pay attention to the severity rankings. Thus, if several failure modes have the same risk priority number, that failure mode with the highest severity should be given more priority. If severity number is the same, those failures with higher occurrence should be given more priority and so on.

Failure Mode & Effect Analysis FMEA

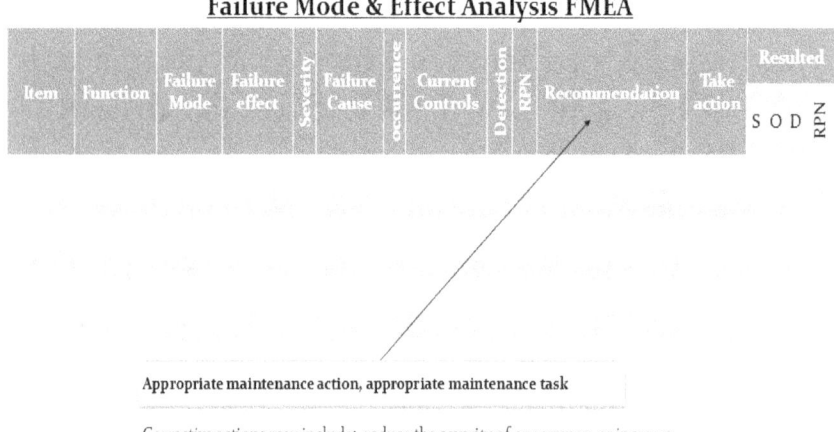

Appropriate maintenance action, appropriate maintenance task

Corrective actions may include: reduce the severity of occurrence ,or increase the detection probability

Step.11 Take Actions to Eliminate or Reduce the High-Risk Failure Modes

This is organized using the problems-solving approaches and implement actions to reduce or eliminate the high-risk failure modes.

Often the easiest way to make an improvement to the product or process is to increase the detectability of the failure, thus lowering the detection rate.

Increase the detection rate can be done though assigning a schedule PM action, use a proper condition monitoring program or consider a mistake proofing method in the design. For example, ac computer software will automatically warn in case of low disk space.

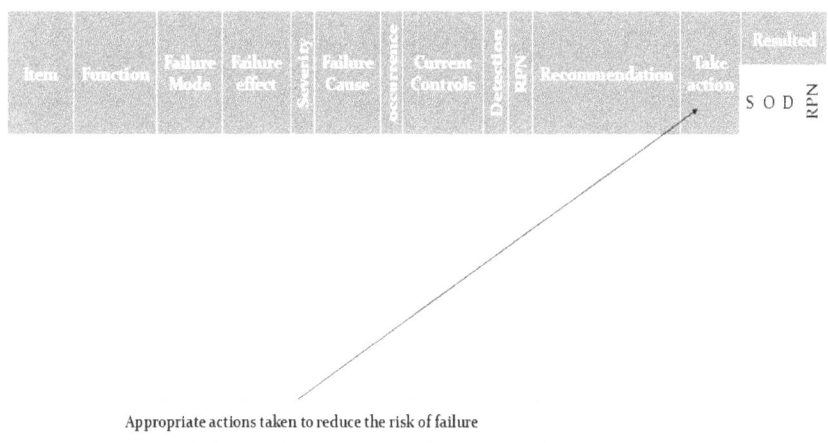

Appropriate actions taken to reduce the risk of failure

Step.12 Calculate the Risk Priority Number RPN as the High Risk is Removed

When moves have been made to lessen the danger need number, another positioning for the seriousness, event, and discovery ought to be determined. What's more, a subsequent RPN is determined. Desire is in any event 50 rate decrease in RPN with the FMEA approach.

There will consistently be a potential for failure modes to happen. The inquiry the organization must pose is how much relative danger the group is eager to take. That answer may depend on the business and the reality of the disappointment. For instance, in the atomic business, there is a little edge for mistakes, they can't hazard a calamity happening. In different enterprises, it might be worthy to face the high challenge.

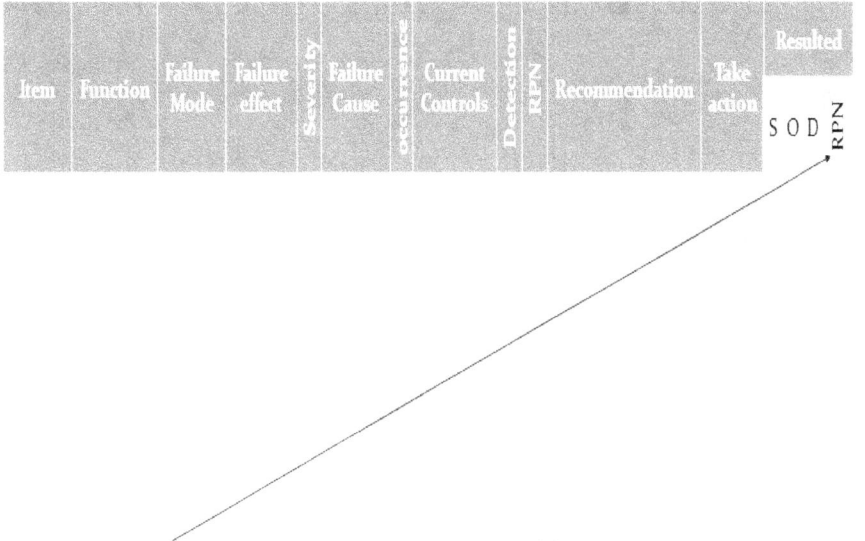

NEW RPN based on the new Severity, Occurrence, and Detection rankings

Examples: Failure Modes, Effects, and Causes

To clarify more, giving some examples of failure modes for some popular components, machines and systems.

Ex.1

Failure mode	Failure Effect	Failure Effect (System)	Failure Effect (End)	Failure cause Level 1	Root cause
Fan operate with high vibration level	Equipment damage/breakdown	Unexpected plant shutdown	Major production losses	Bearing fails	Poor Maint
Fan operate with high vibration level	Equipment damage/breakdown	Unexpected plant shutdown	Major production losses	Housing wear	Poor Maint
Fan operate with high vibration level	Equipment damage/breakdown	Unexpected plant shutdown	Major production losses	Unbalance fan blade	Poor Maint
Fan operate with high vibration level	Equipment damage/breakdown	Unexpected plant shutdown	Major production losses	Looseness in foundation	Poor Maint
Fan operate with high vibration level	Equipment damage/breakdown	Unexpected plant shutdown	Major production losses	Shaft wear	Poor Maint

Ex.2 Transformer

Item name	Failure mode	Failure Effect (local)	Failure Effect (System)	Failure cause	Failure Cause	Root cause
		Functional stop	Production losses	Particles in the oil	Overheated	Bad Maintenance
Oil	1.Short circuit in transformer	Functional stop	Production losses	Water in the oil	Overheated Aging	Bad Maintenance
Tap Changes	2-Can't change voltage level	Functional stop	Production losses	Mechanical damage	Wear	Life time/ maintenance

Ex.3 Water System

Function	Functional failure/failure modes	Causes
Provide water to the industrial process	Total loss of pressure, volume & flow	Pump failed Motor failed Valve out of position

Electric Motor

Function	Functional failure/failure modes	Causes
Drive the water pump	Burn out	Circuit Breaker tripped Bearing seized Insulation Rotor Insulation Stator

Motor Bearing

Failure mode	Failure Cause	Sources of failure/causes	Causes
Bearing seized, this include bearing, seals, lubrication	Lubrication	Contamination	Supply dirty Sealing failed
		Wrong type	Procedure wrong Supply information wrong
		Tool little	Human error Procedure error
		Too much	Human error Procedure error

Final Table

Failure effect			Severity			Causes	Root Cause	Occurrence	Current fault detection methods	Detection	RPN	Actions
Local	sys	end	S	A	C	Seal failed	Seal failed					
Motor shutdown	System shutdown	TPL				Procedure wrong	Lack of training					
						Human error						
						Human error						

Risk=Probability x Severity

Probability or frequency	Consequence or Severity		
	(1) Low	(2) Medium	(3) High
(1) Low	1 L	2 L	3 M
(2) Medium	2 L	4 M	6 H
(3) High	3 M	6 H	9 H

It's important to design your own matrix

FMEA Case Study: A Case of Reliable Improvement:

H ow to Minimize the Risk of Failure for an Electric Transformer by Improving the Service Reliability

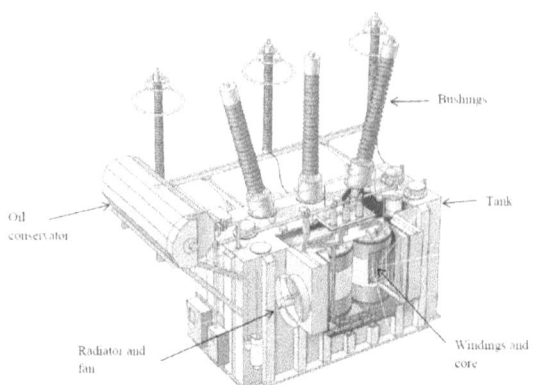

Electric Distribution Transformer for Glass Furnace

Equipment/System Information
Equipment Type: Distribution Transformer.
Technical Specs: 11KV, 2.5KV
Function: Transform electric voltage from 11KV to 400V.
System: Electric station- Supply Glass Furnaces.
Availability of standby system: Generators that provide partial productivity.
Working intervals: 1-2 seconds.
Effectiveness: Avoid furnace damage, but medium **productivity.**

	Severity	Occurrence	Detection	RPN
Initial	7	8	5	280
Revised	7	6	4	168
% Reduction in RPN				40%

RPN Reduction %=Ri-Rr/Ri

The electric transformer is considered critical because a failure causes high production losses – $5,000 an hour. A standby generator could keep the furnace running if the transformer failed. The standby was sufficient to avoid damaging the furnace but did not supply enough electricity to continue production.

Transformer Fault Tree

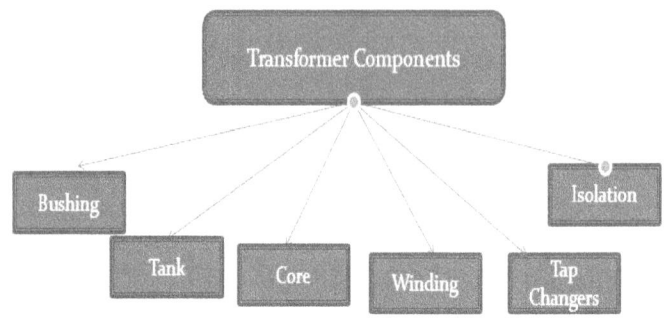

Current Control/Prevention methods

PM type	Component/Item	PM Level
Visual inspection	Oil level	Monthly
	Silica gel	Monthly
	Cooling fans	Monthly
	Temp & gauges	Monthly
Cleaning	External body of the transformer	Monthly
Tightening	Cables	Monthly
Measurements	Voltage	Semi annual
	Ampere	Semi annual
Sampling	Oil	Annually

Failure Log History
Working condition= 24 hrs.

Failure Type	Frequency per 8760 hrs
Oil Heated	3
Short Circuit	2

Volt regulation function error (faulty tap changers)	3

Analysis

Component Name & Function: Bushing, supply high voltage

Failure Mode	Failure Effect	Severity	Failure Causes	Failure Cause	Failure Causes	Failure Cause	Occurrence	Current control detection/prevention methods	Detection	RPN
Short circuit	Equipment shutdown	4	Fault in insulation material	Water penetration or dirt	Inelastic gasket	Aging	1	Visual inspection and cleaning	6	6
				Lack of maintenance			1		6	24
			Damage bushing	Sabotage stone, crash or Careless handling			1		4	16

Recommendation	Take actions	Result			
		S	O	D	RPN
Increase inspection & detectability	Use infrared camera & ultrasound for high detection ability	4	1	2	8
		4	1	2	8

 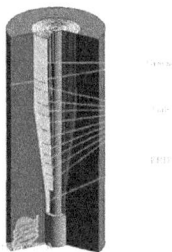

The function of the bushings is to isolate electrical between tank and windings and to connect the windings to the power system outside the transformer.

Component Name & Function: Tank , enclose oil , protect active parts

Failure Mode	Failure Effect	Severity	Failure cause	Failure Cause	Failure Cause	Failure Cause	Occurrence	Current controls	Detection	RPM
Leakage	Equipment shutdown	4	Material/method	Inelastic gasket or corrosion	Aging		1	Visual inspection	5	20
						Insufficient maintenance	1		5	20
			Tank Damage (Rupture)	Mechanical damage	High pressure due to gas generation	Arcing	1	None	10	40
					Careless handling		1		1	14

Recommendation	Take actions	Result			
		S	O	D	RPN
Increase inspection & detectability	Use ultrasound for detection of arcing phenomena	4	1	1	4
		4	1	1	4
		4	1	1	4

The tank is primarily the container of the oil and a physical protection for the active part of the transformer. It also serves as support structure for accessories and control equipment. The tank has to withstand environmental stresses, such as corrosive atmosphere, high humidity and sun radiation. The tank should be inspected for oil leaks, excessive corrosion, dents, and other signs of rough handling.

Component Name & Function: Core, carry magnetic flux

Failure Mode	Failure Effect	Severity	Failure Cause	Failure Cause	Occurrence	Current Control	Detection	RPN
Loss of efficiency (reduction of transformer efficiency)	Lower voltage, production disturbance	4	Mechanical failure	DC magnetization	1	Basic measurements	4	16
				Displacement of the core steal during construction (construction fault)	1		4	16

RPN=S x O x D=16

No Recommendation or actions will be taken here.

48

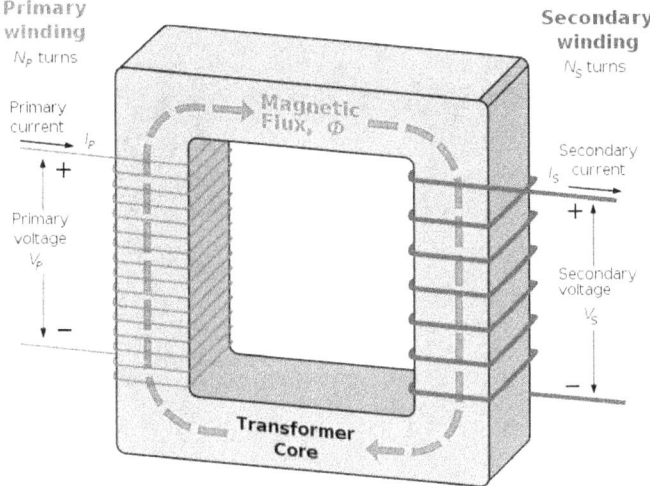

Component Name & Function: Winding, carry current

Failure Mode	Failure Effect	Severity	Failure cause	Failure Cause	Failure Cause	Occurrence	Current Controls	Detection	RPN
Short circuit	Equipment shutdown	4	Fault insulation	Generation of copper sulfide		1		8	32
					Low oil quality	1	Oil sampling	1	4
				Hot spot	Ageing of cellulose	1		5	20
				Movement of transformer	Short circuit in the net	1		5	20
			Mechanical damage	Transient overvoltage	Connection of transformer	1	None	5	20
				Construction fault	Lightning	1		5	20

Recommendation	Take actions	Result			
		S	O	D	RPN
		4	1	2	8
		4	1	2	8
	Use ultrasound for detection of winding problems	4	1	2	8
		4	1	2	8
Increase inspection & detectability		4	1	2	8
		4	1	2	8

51

The windings belong to the active part of a transformer, and their function is to carry current. The windings are arranged as cylindrical shells around the core limb, where each strand is wrapped with insulation paper. Copper is today the primary choice as winding material. In addition to dielectric stresses and thermal requirements the windings have to withstand mechanical forces that may cause windings replacement. Such forces can appear during short circuits, lightning, and short circuits in the net or during a movement of the transformer.

Component Name & function: Oil, the oil serves as both cooling medium and part of the insulation system

Failure Mode	Failure Effect	Severity	Failure cause	Failure Case	Failure Cause	Failure Case	Occurrence	Current Controls	Detection	RPN
Oil	Equipment shutdown	4	Short circuit in transformer	Particles in the oil	Overheated	Pump failure, Dirty particles in the oil	2	Visual monitoring of gauges and oil sampling	4	32
				Water in the oil	Overheated or aging					
			Overheated	Oil is not cooled	Oil, circulation out of function,	Fan/Pump failure	2		4	32
				Overheated	Air/Water cooling is out of function					

Recommendation	Take actions	Result			
		S	O	D	RPN
Increase oil sampling frequency	1. Sample oil every 6 months	4	1	2	8
	2. Increase detectability with infrared camera inspection	4	1	2	8

The transformer oil is a highly refined product from mineral crude oil and consists of hydrocarbon

composition of which the most common are paraffin, naphthenic, and aromatic oils. The oil serves as both cooling medium and part of the insulation system.
The quality of the oil greatly affects the insulation and cooling properties of the transformer. The major causes of oil deterioration are due to moisture and oxygen coupled with heat.
Another function of the oil is to impregnate the cellulose and isolate between the different parts in the transformer.

Component Name & Function : Tap Changers, regulate volt levelling

Function	Failure Mode	Failure Effect	Severity	Failure cause	Failure Cause	Failure Cause	Occurrence	Current Controls	Detection	RPN
Regulate volt leveling	Tap Changes	Change of the voltage output	3	Can't change voltage level	Mechanical damage	Wear	2	Voltage measuring	6	36

Recommendation	Take actions		Result			
		S	O	D	RPN	
Increase inspection & detectability	Use infrared inspection to detect tap changers faults	3	1	2	8	

Motorized Taps ⇒

The function of on-load tap-changer (OLTC) is to regulate the voltage level by adding or subtracting turns from the transformer windings.

Component Name & Function: Solid Isolation, is cellulose based products such as press board and paper. Its function is to provide dielectric and mechanical isolation to the windings.

Failure Mode	Failure Effect	Severity	Failure cause	Sources of failure	Failure Cause	Occurrence	Current Controls	Detection	RPN
Can't supply insulation	Equipment Shutdown	4	Mechanical damage	Short circuit, Movement of transformer	Ageing of cellulose	1		10	40
			fault in insulation material	Ageing of cellulose	Low oil quality, or Overload	1	None	1	4
				Hot spot				10	40
				Generation of copper sulfide		1		10	40

58

Recommendation	Take actions	Result			
		S	O	D	RPN
Increase inspection & detectability	Use ultrasound for detection of winding problems	4	1	2	8
		4	1	2	8
		4	1	2	8

The solid insulation in a transformer is cellulose based products such as press board and paper. Its function is to provide dielectric and mechanical isolation to the windings.

RPN Analysis for Transformer Components

Part/Item	RPN		Part/Item	RPN
	16			20
Bushing	16		Winding	20
	16			20
	20		Oil	32
Tank	20			32
	40		Tap Changers	36
	4			40
Core	16			40
	16		Solid Insulation	4
	4			40
Winding	4		**Total**	**492**
	20			

A cutoff point of RPN 16 can be set because over 50% of the failure modes are above this number.

Total Risk Priority Number= 492.

Recommendations

Increase the detection probability for the following failures:

  -Winding insulation.

  -Tap changers.

  -Oil condition.

  -Insulation breakage.

  -Bushing insulation failure.

  -Tank corrosion/leakage.

Fit more generators to avoid production losses upon transformer failure (we will need more especially if the whole furnaces are working).

Corrective Actions (stage 1):

Usage of thermal camera to monitor the winding, tap changers, oil temp, insulation, bushing and tank corrosion.

Increase visual inspection capability for the tank.

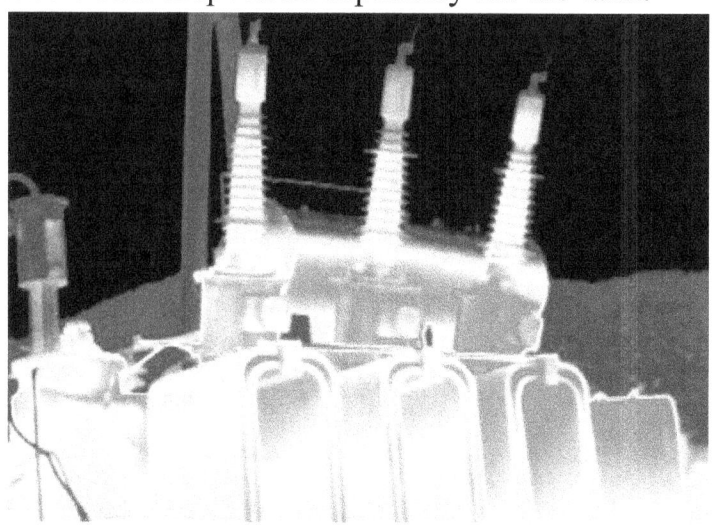

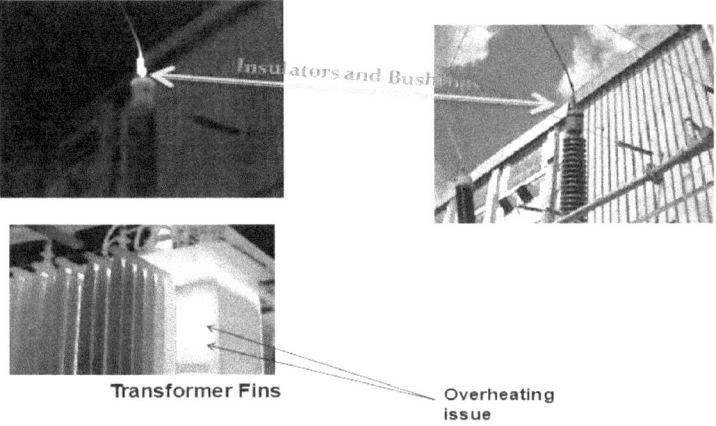

Transformer Fins — Overheating issue

Expected Total Risk Priority Number after applying the corrective actions.

Corrective Actions (stage 2):

Use the Ultrasound detection to detect winding problems & isolation.

Expected Total Risk Priority Number after applying the corrective actions (stage 1 &2): Supportive for early detection.

RPN Reduction %=R initial – R revised/

=492-184/492.

=62%

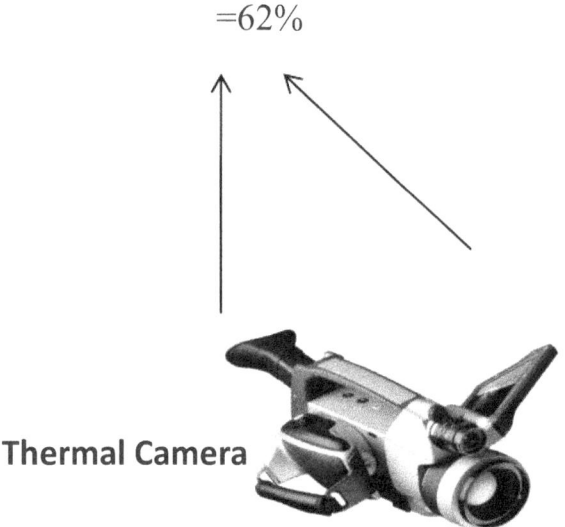

Thermal Camera

Increase inspection reduce the risk of failure.

The improvements that yielded success included using ultrasound to detect issues, increasing the frequency of oil sampling and using infrared analysis to detect mechanical damage.

Detect Transformer Problems

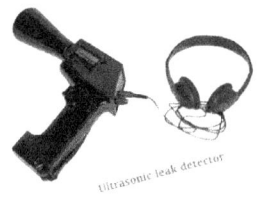

ultrasonic leak detector

Electric Discharges:
•Arcing
•Corona
•Tracking

Remember FMEA is a Team Work Job! Team Members for FMEA:

- Process Engineer.
- Operators.
- Quality.
- Safety.
- Maintenance.
- Product engineer.
- Customer.
- Supplier.

About Transformer Case Study:

This is one example for a component exist in a productivity system, where failure can have negative effect on productivity, maintenance and therefore costs. There are more critical components that failure can effect safety and they need more attention. When breaking down the system and using FMEA, a lot of compoenents will have high numbers of severity. If one of these components are more likely to fail (or have failed before) and there is no propoer prevention/control method cureently applied, the RPN for this compoennt will be big, it should be given more priority and tackled first.

High critical equipment or components exit in critical system in general should be all enrolled in FMEA process.

FMEA and Continuous Improvement

E ach step is a FMEA toward the target, so please repeat the FMEA wheel!

An FMEA process can trigger a number of such actions to improve a product's service or maintenance processes. They include, but are not limited to:

- ✓ Increase the detection rate of high-risk failures using a proper technique to monitor conditions.
- ✓ Increase the inspection rate for a specific component or part.
- ✓ Modify the routine maintenance program.
- ✓ Increase the frequency of replacing a specific spare part.
- ✓ Modify the preventive maintenance schedule.
- ✓ Change a spare part supplier.
- ✓ Redesign a specific part in the system – or redesign the whole system.
- ✓ Use different types of materials or spare parts.

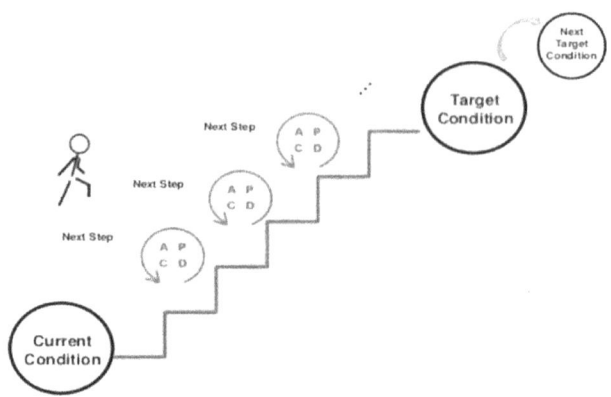

FMEA wheel acts the same as PDCA wheel

Does FMEA Sound Like a Standalone Tool?

Failure mode and effects analysis work with the other Quality tools to maximize a product's reliability. It doesn't work as a standalone tool. For example, to determine occurrence ratings, FMEAs rely on the failure log history, and the documentation process also is important. Problem-solving techniques like "five whys," brainstorming, fault-tree analysis and Pareto analysis must be engaged. These techniques will help determine potential failure modes; assign the severity, occurrence and detection rankings; and provide solutions or actions to eliminate those failures.

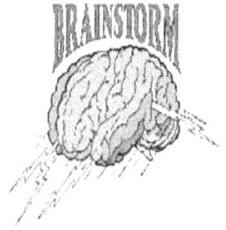

How to get the why?

= Go and see =

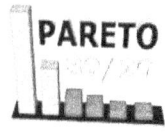

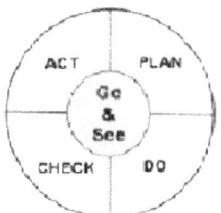

✓ Remember to brainstorm potential failure modes.
✓ Use Pareto to prioritize actions, and which failure mode to tackle first. It doesn't make sense to make improvements to all failures at once! This will increase project cost and results may not be seen to managers as cost effective.
✓ Use 5Whys to solve problems and find the real root cause of issues.
✓ Remember Plan-Do-Check-Act cycle is an endless cycle. You will never be perfect until repeating it again and again.

References:

Raymond J. Mikulak, Robin McDermott. (2008). The Basics of FMEA. Productivity Press; 2nd edition.

Robert T. Amsden and Davida M. Amsdenand. (1998). SPC Simpliefied: Practical steps to quality. Productivity Press; 2 edition.

Soliman, M.H.A. 2014. Analyzing Failure to Prevent Problems. Industrial Management.

Soliman, M.H.A. 2020. Industrial Applications of Infrared Thermography. KDP and Lulu Press.

Soliman, M.H.A. 2020. Ultrasound Analysis for Condition Monitoring. KDP.

KTH Electrical Engineering.

About the Author

 Mohammed Hamed Ahmed Soliman is an industrial engineer, consultant, university lecturer, operational excellence leader, and author. He works as a lecturer at the American University in Cairo and as a consultant for several international industrial organizations.

Soliman earned a bachelor of science in Engineering and a master's degree in Quality Management. He earned post-graduate degrees in Industrial Engineering and Engineering Management. He holds numerous certificates in management, industry, quality, and cost engineering.

For most of his career, Soliman worked as a regular employee for various industrial sectors. This included crystal-glass making, fertilizers, and chemicals. He did this while educating people about the culture of continuous improvement.

Soliman has lectured at Princess Noura University and trained the maintenance team in Vale Oman Pelletizing Company. He has been lecturing at The American University in Cairo for 6 year and has designed and delivered 40 leadership and technical skills enhancement training modules.

Soliman is a member at the Institute of Industrial and Systems Engineers and a member with the Society for Engineering and Management Systems. He has published several articles in peer reviewed academic journals and magazines. His writings on lean manufacturing, leadership, productivity, and business appear in Industrial Engineers, Lean Thinking, and Industrial Management. Soliman's blog is www.personal-lean.org.

Recommended Reads by Soliman:

www.ingramcontent.com/pod-product-compliance
Lightning Source LLC
Chambersburg PA
CBHW070510220526
45467CB00002B/614